T0225224

Cambridge Elements ☰

Elements in Geochemical Tracers in Earth System Science
edited by
Timothy Lyons
University of California, Riverside
Alexandra Turchyn
University of Cambridge
Chris Reinhard
Georgia Institute of Technology

THE URANIUM ISOTOPE PALEOREDOX PROXY

Kimberly V. Lau
University of California, Riverside and University of Wyoming

Stephen J. Romaniello
Arizona State University and University of Tennessee, Knoxville

Feifei Zhang
Yale University

CAMBRIDGE
UNIVERSITY PRESS

CAMBRIDGE
UNIVERSITY PRESS

University Printing House, Cambridge CB2 8BS, United Kingdom

One Liberty Plaza, 20th Floor, New York, NY 10006, USA

477 Williamstown Road, Port Melbourne, VIC 3207, Australia

314–321, 3rd Floor, Plot 3, Splendor Forum, Jasola District Centre, New Delhi – 110025, India

79 Anson Road, #06–04/06, Singapore 079906

Cambridge University Press is part of the University of Cambridge.

It furthers the University's mission by disseminating knowledge in the pursuit of education, learning, and research at the highest international levels of excellence.

www.cambridge.org
Information on this title: www.cambridge.org/9781108731119
DOI: 10.1017/9781108584142

First published 2019

A catalogue record for this publication is available from the British Library.

ISBN 978-1-108-73111-9 Paperback
ISSN 2515-7027 (online)
ISSN 2515-6454 (print)

The Uranium Isotope Paleoredox Proxy

Elements in Geochemical Tracers in Earth System Science

DOI: 10.1017/9781108584142
First published online: August 2019

Kimberly V. Lau
University of California, Riverside and University of Wyoming

Stephen J. Romaniello
Arizona State University and University of Tennessee, Knoxville

Feifei Zhang
Yale University

Author for correspondence: Kimberly V. Lau, kimberly.lau@uwyo.edu

Abstract: Uranium isotopes ($^{238}U/^{235}U$) have emerged as a proxy for reconstructing the redox conditions of the Earth's oceans and atmosphere based upon the large isotopic fractionation between reduced U(IV) and oxidized U(VI). Variations in $^{238}U/^{235}U$, particularly when recorded in carbonate sediments, can track global trends in marine oxygenation and deoxygenation. It is unique among proxies because reduction occurs primarily at the sediment–water interface, and this sensitivity makes U isotopes especially relevant for the habitability of benthic animals. This Element covers the background, methods, and case studies of this promising tool for understanding Earth's environmental transitions, as rapid development continues to refine the accuracy of interpretations of $^{238}U/^{235}U$ records.

Keywords: uranium isotopes, marine anoxia, paleoredox, carbonates

ISBNs: 9781108731119 (PB), 9781108584142 (OC)
ISSNs: 2515-7027 (online), 2515-6454 (print)

Contents

1 Introduction

Uranium is naturally redox sensitive: its solubility varies between its two main oxidation states, where U(VI) is soluble and U(IV) is insoluble. The intermediate U(V) is sparingly present and undergoes rapid disproportionation into U(IV) and U(VI). Because the solubility of U in natural waters is dependent on its oxidation state, U concentrations ([U]) have been used to track global marine redox conditions using chemical sediments, such as calcium carbonate precipitates (e.g., Russell et al., 1994). Such interpretations, however, are not straightforward because [U] reflect local depositional factors in addition to the size of the seawater U reservoir. Specifically, the distribution coefficient of U is sensitive to the original $CaCO_3$ polymorph as well as U speciation, which is pH dependent (Meece and Benninger, 1993; Russell et al., 2004).

Additional constraints on seawater U may be obtained through its isotopic ratios. Uranium has three naturally occurring radioactive isotopes: ^{238}U (99.28%), ^{235}U (0.72%), and ^{234}U (0.0054%), with half-lives of ~4.46 × 10^9, ~704 × 10^6, and ~245 × 10^3 years, respectively. Because ^{234}U has a short half-life and is readily expelled from minerals via alpha recoil during the decay of ^{238}U, U-series disequilibria and the $^{234}U/^{238}U$ ratio have been used to track the rate and timescale of oceanographic processes on 1–800 kyr time scales (e.g., Henderson and Anderson, 2003). Because uranium is the heaviest natural element, mass-dependent isotopic fractionation of U is relatively small, and fractionation is instead dominated by mass-independent effects related to relative differences in the size and shape of odd- and even-mass isotopes. This anomalous pattern of isotopic behavior was first observed in redox ion-exchange experiments (Fujii et al., 1989; Nomura et al., 1996). Bigeleisen (1996) subsequently provided a theoretical description for this nuclear volume effect, and these calculations were later supported by the work of Schauble (2006) and Abe et al. (2008). This isotopic fractionation was observed in natural environmental samples by Stirling et al. (2007) and Weyer et al. (2008), with results expressed in delta notation as:

$$\delta^{238}U = \left(\frac{^{238}U}{^{235}U}_{sample} \Big/ \frac{^{238}U}{^{235}U}_{standard} - 1 \right) \times 1000 \qquad (1.1)$$

Although the absolute value of $^{238}U/^{235}U$ ratios varies with time due to radioactive decay, values expressed as $\delta^{238}U$ are time invariant because ^{238}U and ^{235}U undergo decay in samples and the reference material at the same rate. Because of this, a 1‰ fractionation of $^{238}U/^{235}U$ in Archean-age sediments is still preserved as a 1‰ value measured in the laboratory today.

The first effort to use $\delta^{238}U$ to track redox conditions was in black shales deposited during oceanic anoxic event 2 (OAE2), a period of ocean deoxygenation at the Cenomanian–Turonian boundary (~94 Ma; Montoya-Pino et al., 2010). This was followed by measurements of $\delta^{238}U$ in shallow-marine carbonate rocks, with the first record from the end-Permian mass extinction (~252 Ma) suggesting that the onset of ocean anoxia was concurrent with the extinction (Brennecka et al., 2011b). This record has since been extended from the Middle Permian through the Middle Triassic, and the observed trends in $\delta^{238}U$ have been replicated in multiple locations across oceanic basins and depositional environments (Lau et al., 2016; Elrick et al., 2017; Zhang et al., 2018a, 2018b, 2018c). At the time of publication, 25 $\delta^{238}U$ papers have been published that span the last 4 Ga (Kendall et al., 2013; Wang et al., 2018), with 16 focused on the carbonate rock record. Combined, these efforts have helped to elucidate how the redox conditions of the oceans and atmosphere have evolved through time.

Other commonly used inorganic geochemical paleoredox proxies include enrichments of other redox-sensitive elements and their isotopes. Popular proxies include iron speciation; rhenium (Re), vanadium (V), and molybdenum (Mo) concentrations; and rare earth element (REE) enrichments. The list of isotopic proxies is growing and includes iron (Fe), chromium (Cr), molybdenum (Mo), and sulfur (S) isotopes. Uranium is unique among these proxies for the following reasons. First, its redox potential is similar to iron reduction – requiring an absence of oxygen but not euxinic (i.e., sulfidic) conditions – meaning that it potentially can capture conditions that are particularly relevant for typical marine macrofauna. In contrast, iron speciation can track the prevalence of Fe^{2+} versus H_2S, and Mo is representative of H_2S concentrations. Second, its mode of reduction is unique among the elements. For example, Mo and V are generally quantitatively removed from euxinic water columns, meaning that their isotopic ratios directly reflect water column compositions. In contrast, water-column U is not quantitatively reduced in any modern setting, resulting in variable offsets between seawater and reducing settings that complicate interpretations of $\delta^{238}U$.

Third, most or all U reduction occurs at the sediment–water interface rather than in the water column. Although U reduction is thermodynamically predicted in modern anoxic water columns, U(IV) has not been observed in the Black Sea water column, possibly because of limited mineral surfaces or because reduction is inhibited by aqueous complexation of U(VI) (Anderson et al., 1989). Thus, U is inferred to be a tracer of bottom-water anoxia – instead of water-column conditions – which is important for the habitability of benthic and epifaunal marine environments. Sedimentary U cycling can be decoupled from water-column conditions if organic carbon burial rates are high, stimulating U reduction in sediments (McManus et al., 2005; Morford et al., 2009). Fourth, U has a long residence time in the ocean (~450 kyr today; Ku et al., 1977), and thus acts as a conservative element in contrast to many other elemental and isotopic tracers. This is supported by uniform $\delta^{238}U$ values in the modern ocean (summarized in Andersen et al., 2016). In sum, if it is possible to reconstruct ancient seawater $\delta^{238}U$, it may be possible to quantitatively estimate the global area of anoxic seafloor that is linked to U reduction.

As a global proxy, $\delta^{238}U$ provides unique information that cannot be inferred from concentrations alone, which are sensitive to factors such as sedimentation rate and mineralogy, or from explicitly local proxies such as iron speciation and REE anomalies, which reflect only the local depositional environment. As described in detail below, $\delta^{238}U$ can also be tracked using carbonate rocks, whereas many other proxies are limited to organic-rich shales. For these reasons, despite their recent arrival in the paleoredox toolkit, U isotopes have expanded in popularity and potential as a tracer of redox conditions through Earth history.

2 Systematics of the Uranium Isotope Paleoredox Proxy

The isotopic fractionation resulting from the nuclear volume effect is estimated to range from 0.7 to 1.4‰ ($\Delta = \delta^{238}U_{U(IV)} - \delta^{238}U_{U(IV)}$; e.g., Schauble, 2006; Abe et al., 2008; Stylo et al., 2015; Brown et al., 2018) with preferential removal of ^{238}U into reducing sediments. The $\delta^{238}U$ value of U in modern seawater (-0.39 ± 0.01‰; 2 S.D.) is controlled by the sources and sinks of U to the ocean (Figure 1; e.g., Weyer et al., 2008; Tissot and Dauphas, 2015; Andersen et al., 2016, 2017). The major input flux is riverine delivery of weathered U with a mean $\delta^{238}U$ of -0.26 ± 0.32‰ (1 S.D.) and a flux-normalized value of -0.34‰ (Andersen et al., 2016), which generally reflects the bulk earth $\delta^{238}U$ of -0.31 ± 0.03‰ (Andersen et al., 2015) and bulk continental crust of -0.28 to -0.30 ± 0.03‰ (Tissot and Dauphas, 2015). Minor input fluxes include dust and submarine groundwater discharge, where the $\delta^{238}U$ of dust is expected to be similar to crustal values and no constraints have been published for the latter.

There are numerous processes that remove U from the ocean. Their relative proportions, calculated from the average values of published budgets (Barnes and Cochran, 1990; Morford and Emerson, 1999; Dunk et al., 2002; Henderson and Anderson, 2003), and their isotope values are shown in Figure 1. Volumetrically, the most important flux is removal to sediments deposited in reducing/hypoxic settings (also referred to as the suboxic flux), which

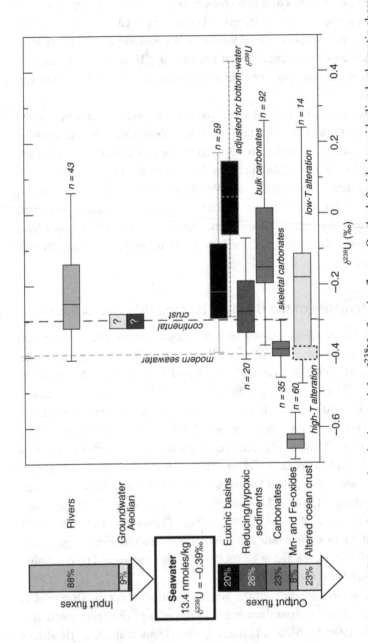

Figure 1 Overview of the modern uranium budget and the δ^{238}U of major fluxes. On the left side is an idealized schematic showing the major input and output fluxes of uranium to the ocean, with the height of the box sized to its relative proportion (Barnes and Cochran, 1990;

comprises ~26% of the total U output and has a fractionation estimated at ~0.15‰ (Weyer et al., 2008; Andersen et al., 2016). The most isotopically distinct flux is removal of U(IV) via sediments deposited in euxinic settings, which represents ~20% of U removal in total and has the most positive $\delta^{238}U$ value. The fractionation from seawater into euxinic sediments has been proposed to be 0.5 to 0.7‰ from modern observations (Andersen et al., 2014; Holmden et al., 2015; Noordmann et al., 2015; Rolison et al., 2017). Because most U reduction occurs in sediments, this observed (or effective) fractionation differs from the predicted (or intrinsic) fractionation because of diffusion limitation of U(VI) into shallow sediments (Clark and Johnson, 2008; Andersen et al., 2014).

Oxidized U can be incorporated into carbonate rocks, removing ~23% of seawater U in total with a fractionation described in detail in Section 3.1.

Caption for Figure 1 (cont.)

Morford and Emerson, 1999; Dunk et al., 2002; Henderson and Anderson, 2003). To the right are box-and-whisker plots of $\delta^{238}U$ data for each flux, with the range, median, and quartile values indicated. Riverine data are from Stirling et al. (2007), Noordmann et al. (2016), Tissot et al. (2015), and Andersen et al. (2016). Groundwater and aeolian $\delta^{238}U$ may represent bulk continental crust, which has a value of –0.29‰ (Tissot et al., 2015). A selection of euxinic basin $\delta^{238}U$ are plotted, from sites for which bottom-water $\delta^{238}U$ is known. These include Kyallren Fjord and Landsort Deep (Noordmann et al., 2015), Saanich Inlet (Holmden et al., 2015), and the Black Sea (Rolison et al., 2017). Bottom-water $\delta^{238}U$ data were not measured by Weyer et al. (2008), Montoya-Pino et al. (2010), and Andersen et al. (2014), and so these data are not shown. Because waters in modern euxinic basins are more negative than the open ocean due to reservoir drawdown, we also show the relative offset of sediments in this sink relative to bottom-water $\delta^{238}U$ in the dashed box. Reducing/hypoxic sediments are from high-productivity regions from the Washington Coast (Andersen et al., 2016) and the Peru Margin (Weyer et al., 2008). Bulk carbonate sediments less than 1 Ma in age are from Romaniello et al. (2013), Chen et al. (2018a), and Tissot et al. (2018). We also show $\delta^{238}U$ of modern skeletal carbonates from Stirling et al. (2007), Weyer et al. (2008), Andersen et al. (2014), and Chen et al. (2018b). The $\delta^{238}U$ of Fe and Mn oxides (ferromanganese crusts) are from Goto et al. (2014) and samples <1 Ma from Wang et al. (2016). Low-temperature alteration data are from Andersen et al. (2015); it is assumed that no fractionation occurs during exchange with basalt at high temperature (dashed box).

Additionally, U(VI) can be adsorbed onto ferromanganese crusts – a minor flux representing 8% of U removal but with a fractionation of –0.24‰ due to a change in coordination state during sorption to Mn oxides (Brennecka et al., 2011a; Goto et al., 2014; Wang et al., 2016). Incorporation into Fe oxides, such as banded iron formations or ironstones, is included with this flux. Only a few experiments have explored isotopic fractionation during sorption to Fe oxides, showing relatively little fractionation (Stylo et al., 2015). Finally, a large proportion of U is removed via basalt alteration (or altered ocean crust). Although low-temperature alteration imparts a measurable fractionation, it comprises a smaller proportion of the total flux than does high-temperature alteration, which is assumed to have no fractionation (Andersen et al., 2015; Noordmann et al., 2016). Burial of U into oxic sediments (such as hemipelagic red clays) may impact the U budget but is poorly constrained and warrants further study. Fractionation under ferruginous conditions (anoxic but not euxinic) is also not well understood.

In sum, theoretical calculations, experimental results, and observations of geologic materials indicate that the largest fractionation of $^{238}U/^{235}U$ occurs between U(VI) and U(IV), leading to the inference that isotopic fractionation occurs during reduction (Schauble, 2006; Stirling et al., 2007; Weyer et al., 2008; Stylo et al., 2015). Thus, $\delta^{238}U$ has been proposed to track U reduction, and hence, environmental redox conditions. Fractionation has been suggested to occur only during microbial reduction and exchange of U(VI) and U(IV) (Stylo et al., 2015); however, recent experiments indicate that fractionation is possible whether reduction occurs biotically or abiotically (Brown et al., 2018). Based on this framework, more extensive anoxia – globally integrated – is expected to result in more negative seawater $\delta^{238}U$, which could then be tracked in sedimentary rock archives.

3 Materials and Methods

The $\delta^{238}U$ paleoredox proxy has been measured in various chemical and siliciclastic sediments including carbonate rocks, organic-rich shales, Mn oxide crusts, and ironstones. Like any geochemical proxy, interpretations should be supported by analyzing multiple stratigraphic sections, preferably in distinct paleogeographical locations and/or across depositional environments (Lau et al., 2016; Clarkson et al., 2018; Zhang et al., 2018b). Moreover, coupling $\delta^{238}U$ records across multiple types of lithologic archives may help to characterize seawater $\delta^{238}U$ while accounting for carbonate diagenesis and local controls on U fractionation.

3.1 Carbonate Rocks

Shallow marine carbonate rocks are ideal archives because they have been deposited continuously since ca. 3.8 Ga and are often preserved in multiple oceanic basins. Primarily comprised of calcite/aragonite ($CaCO_3$) and dolomite ($CaMg(CO_3)_2$), their precipitation and preservation are linked with ocean conditions, making them suitable recorders of marine chemistry. If precipitated under geochemical conditions that result in comparable aqueous U speciation, aragonite and calcite should exhibit similar fractionation (Chen et al., 2018b). As discussed in detail in Section 4.2, the reliability of the shallow marine carbonate $\delta^{238}U$ record is supported by the reproducibility of multiple congruent $\delta^{238}U$ records showing the same pattern and absolute values in carbonate sediments deposited during the Late Permian to Early Triassic from multiple locations in the Tethys Sea and the Panthalassic Ocean (Brennecka et al., 2011b; Lau et al., 2016; Elrick et al., 2017; Zhang et al., 2018a, 2018b, 2018c).

The geochemistry of carbonate rocks is susceptible to syn- and postdepositional geochemical alteration. Early diagenesis imparts a positive offset of 0.2 to 0.3‰ relative to seawater, with a mean of 0.24 to 0.27‰ observed in late Neogene shallow-marine carbonate sediments (Romaniello et al., 2013; Chen et al., 2018a; Tissot et al., 2018). The speciation of U, which is strongly related to pH, Ca concentration, ionic strength, and alkalinity, can impart variable isotopic fractionation (Chen et al., 2017). Primary carbonate precipitates such as skeletal grains and ooids record seawater $\delta^{238}U$ with a minor fractionation (0.0 to 0.09‰; Romaniello et al., 2013; Chen et al., 2018b), with subsequent recrystallization, changes in U speciation, and incorporation of U(IV) into carbonate cements all having a role in producing the positive offset in $\delta^{238}U$ (Chen et al., 2017; Chen et al., 2018a; Tissot et al., 2018). Moreover, the microfacies in a carbonate rock may represent different phases of precipitation and recrystallization, potentially resulting in distinct initial and subsequent $\delta^{238}U$ values (Hood et al., 2016, 2018). Nonetheless, existing studies demonstrate that if diagenetic offsets are considered, the $\delta^{238}U$ of marine carbonates can faithfully track seawater patterns.

3.2 Organic-Rich Shales

Organic-rich shales (or black shales) can also be used to track seawater $\delta^{238}U$. Because of their relatively low permeability, shales are generally thought to be less prone to diagenesis. This application relies on the assumption that the isotopic offset between U in seawater and organic-rich sediments is known. The $\delta^{238}U$ of the Middle Devonian Marcellus shale and its interbedded limestone units was used to determine the isotopic fractionation during this time interval and found that it matches estimates for this fractionation in modern settings of 0.6‰ (Phan et al., 2018). In addition to reconstructing anoxic events (such as OAE2; Montoya-Pino et al., 2010), $\delta^{238}U$ data in organic-rich shales have also been used to infer the rise of atmospheric pO_2 (Asael et al., 2013; Kendall et al., 2013; Yang et al., 2017; Wang et al., 2018). This is based on the premise that without atmospheric O_2, the $U(IV)O_2$ mineral uraninite would be weathered from continental rocks, transported as a detrital grain, and buried in sediments with the $\delta^{238}U$ value of bulk Earth. Once atmospheric pO_2 reached the threshold to oxidize uraninite, the seawater $U(VI)$ reservoir increased, permitting U redox-cycling and isotopic fractionation.

Although $\delta^{238}U$ records in organic-rich shales show promise, these sedimentary rocks represent a complex combination of local and global signatures because U does not undergo quantitative water-column reduction. This is further complicated by the fact that organic-rich shales are often deposited in isolated, semirestricted basins. Partial drawdown of U in such environments leads to isotopic Rayleigh distillation effects, resulting in water-column $\delta^{238}U$ that is lower than global average seawater – as is observed in the Black Sea (Rolison et al., 2017). In turn, this results in authigenic $\delta^{238}U$ in Black Sea sediments that is lower than would be expected for highly reducing sediments in direct contact with global average seawater (Andersen et al., 2014). This is discussed in further detail in Section 4.3.

3.3 Other Sedimentary Archives

Alternative $\delta^{238}U$ archives include Mn oxides and iron chemical sediments (ironstones or banded iron formations). Mn oxides can be attractive because they have a well-constrained isotopic fractionation (Brennecka et al., 2011a; Goto et al., 2014) and can provide a long record of seawater $\delta^{238}U$. Surprisingly, the ferromanganese crust record reveals that the long-term seawater $\delta^{238}U$, integrated over ~1 Myr of time for each sample, may have been nearly invariant over the past 80 Ma (Wang et al., 2016) and suggests that redox events lasting over 1 Myr were absent. Interpretations of $\delta^{238}U$ in ironstones – deposited in the Archean and Proterozoic – offer an important glimpse of U cycling on early Earth, although the mechanisms and isotopic fractionation of U incorporation into ironstones are not as well understood as for modern ferromanganese precipitates. Last, the $\delta^{238}U$ of paleosols may fingerprint the presence of mobile $U(VI)$ (Wang et al., 2018). A compilation of $\delta^{238}U$ in ironstones, black shales, and paleosols suggests that a shift in the Mesoarchean may be linked to the onset of uraninite oxidative weathering (Wang et al., 2018).

3.4 Analytical Methods

The $^{238}U/^{235}U$ proxy is commonly measured using multicollector ICP-MS with a double-spike method to correct for instrument mass bias. Ratios are typically reported relative to the CRM-145 or CRM-112a standard (CRM-145 is the solution derived from the CRM-112a metal) with an absolute $^{238}U/^{235}U$ ratio of 137.844 (Condon et al., 2010). Samples are typically spiked using the IRMM-3636 ^{233}U-^{236}U double spike, and U is separated and purified via column chromatography using a U-specific resin (e.g., Eichrom UTEVA^{TM}) following procedures introduced by Stirling et al. (2007) and Weyer et al. (2008). Because U is present as a trace element in sedimentary rocks, sample sizes are relatively large and typically range from 100 to 400 mg for shales (e.g., Kendall et al., 2013; Wang et al., 2018) and from 1 to 9 g for carbonates (e.g., Dahl et al., 2014; Azmy et al., 2015). However, analytical techniques can be improved to allow for small sample sizes (e.g., Wang et al., 2016), and some research groups utilize repeated or additional purification steps to improve the analytical precision for challenging samples (e.g., Chen et al., 2018b).

A challenge specific to carbonates is that the dissolution procedure to extract the geochemical constituents may selectively target detrital, organic, or diagenetic phases that are not representative of the primary or initial $\delta^{238}U$. Carbonates are usually leached to obtain the carbonate fraction and to avoid detrital and organic components by using a weaker acid (e.g., 0.25 to 1 M HCl, 1 M HNO$_3$, 1 M CH$_3$COOH), which varies between labs (reviewed in Tissot et al., 2018). Currently, most methods aim to completely dissolve the carbonate fraction. Although the published data do not indicate that either the acid strength or type have a measurable impact on the analyzed $\delta^{238}U$ values, there is the potential to produce data that are not directly comparable. Digestion of siliciclastic sedimentary rocks is more routine, utilizing an ashing step followed by a multistep acid dissolution procedure (see Asael et al., 2013).

4 Case Studies

4.1 Compilations across Geologic Time

A current compilation of all published $\delta^{238}U$ data from all sedimentary archives is provided in Figure 2. Despite the temporal gaps, the compiled records demonstrate several key points. First, the overall trend shows an increase in $\delta^{238}U$ values over time, with mean values in shale and carbonate rock archives increasing from –0.13‰ ($n = 292$) and –0.56‰ ($n = 205$) prior to 500 Ma to –0.06‰ ($n = 59$) and –0.37‰ ($n = 613$) after 500 Ma, respectively. These averages do not include sediments deposited within the last 3 Ma, and as a result this compilation highlights the lack of Cenozoic $\delta^{238}U$ data. Note that the oldest carbonate data are in the Cryogenian, approximately 2.8 Ga younger than the oldest shale data. Nonetheless, the observed increases in shale and carbonate $\delta^{238}U$ after 500 Ma are significant (Wilcoxon rank sum test p-value = 0.0006

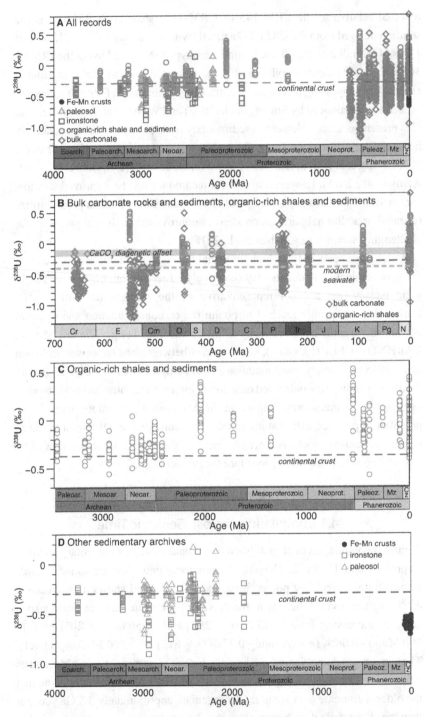

Figure 2 Compilation of $\delta^{238}U$ data. The gray dashed line marks the continental crust $\delta^{238}U$ value. **(A)** $\delta^{238}U$ in bulk carbonate rocks (diamonds), organic-rich

for shales and $< 1 \times 10^{-15}$ for carbonates) and suggest a long-term evolution of the U budget.

Archean $\delta^{238}U$ values from ironstones and organic-rich shales have a mean of $-0.33‰$ (2.6 to 3.75 Ga; $n = 160$) and are similar to bulk Earth ($-0.31‰$; Andersen et al., 2015), which has been interpreted to result from low atmospheric pO_2 and U cycling dominated by detrital U(IV). An increase in atmospheric pO_2 oxidized continental U(IV) minerals such as uraninite, leading to the growth of the seawater U(VI) reservoir—which was then subject to marine redox conditions and attendant isotopic fractionation from preferential enrichment of ^{238}U into the authigenic U(IV) of marine shales relative to seawater, resulting in higher isotopic values (Asael et al., 2013; Kendall et al., 2013; Wang et al., 2018). A U cycle that is sensitive to redox processes is supported by the $\delta^{238}U$ record from 500 Ma through the Mesozoic. The mean $\delta^{238}U$ of organic-rich shales from this time interval is 0.25‰ greater than bulk Earth $\delta^{238}U$. This shift may be attributed to the growth and evolution of predominantly oxygenated sinks, including U(VI) in seawater, which ultimately is buried in carbonates, and incorporation of U(VI) into ferromanganese crusts. In the modern budget, the $-0.24‰$ fractionation associated with this latter sink is sufficient to

---

Caption for Figure 2 (cont.)

shales (open circles), ironstones (squares), paleosols (triangles), and ferromanganese crusts (filled circles) over the last 4 Ga. **(B)** $\delta^{238}U$ from carbonate rocks /sediments and organic-rich shales/sediments, over the last 700 Ma. The diagenetic offset from seawater $\delta^{238}U$ in recent carbonate sediments is shown in the bar (Chen et al., 2018a; Tissot et al., 2018), with modern seawater marked with the lower dashed line. **(C)** $\delta^{238}U$ from organic-rich shales over the last 3.5 Ga. **(D)** $\delta^{238}U$ in other sedimentary archives across the last 4 Ga. Data from ferromanganese crusts are from Goto et al. (2014) and Wang et al. (2016), and ironstone and paleosol data are from Wang et al. (2018). Carbonate data are from Brennecka et al. (2011b), Dahl et al. (2014), Azmy et al. (2015), Lau et al. (2016), Hood et al. (2016), Dahl et al. (2017), Elrick et al. (2017), Jost et al. (2017), Lau et al. (2017), Song et al. (2017), Bartlett et al. (2018), Clarkson et al. (2018), Wei et al. (2018), White et al. (2018), and Zhang et al. (2018a, 2018b, and 2018c). Shale data are from Stirling et al. (2007), Weyer et al. (2008), Montoya-Pino et al. (2010), Asael et al. (2013), Kendall et al. (2013), Kendall et al. (2015), Lu et al. (2017), Yang et al. (2017), Phan et al. (2018), and Wang et al. (2018). Modern reducing sediment data are from Weyer et al. (2008), Montoya-Pino et al. (2010), Andersen et al. (2013), Holmden et al. (2015), Noordmann et al. (2015), Andersen et al. (2016), Hinojosa et al. (2016), and Rolison et al. (2017).

achieve isotopic mass balance, although we note that constraints on the $\delta^{238}U$ of pelagic and hemipelagic clays and deep-sea siliceous and calcareous sediments are limited (Anderson et al., 2015; Tissot and Dauphas, 2015; Andersen et al., 2016).

Second, the range of $\delta^{238}U$ over short time intervals (< 1 My) greatly exceeds the billion-year trend. For example, the $\delta^{238}U$ of modern organic-rich shales covers ~0.8‰ (Figure 1). This is partly a function of variability in reducing sedimentary environments including basin restriction (e.g., Andersen et al., 2014). Third, the majority of studies have concentrated on identifying $\delta^{238}U$ patterns around major, transient perturbations, such as mass extinctions and known oceanic anoxic events. In other words, investigations of $\delta^{238}U$, particularly from the Neoproterozoic to today, have selectively focused on capturing time intervals where large variations in marine redox conditions are expected. These events mostly spanned <1 My – ideal targets for capturing redox-driven seawater $\delta^{238}U$ shifts given the modern residence time of U of ~450 kyr (Ku et al., 1977). Future studies of $\delta^{238}U$ between major events – from periods where fluctuations in ocean anoxia are not expected – will elucidate the range and absolute values of $\delta^{238}U$ during background time intervals.

4.2 The End-Permian Mass Extinction

The Paleozoic–Mesozoic boundary (ca. 252 Ma) experienced the most catastrophic mass extinction event in the Phanerozoic, with over 80% of marine genera disappearing (see references in Schaal et al., 2015). Anoxia has been proposed to play a role in the extinction and the prolonged recovery in biodiversity, which lasted up to five million years. The published $\delta^{238}U$ data over this interval (Figure 3) span depositional environments from the interior of carbonate platforms to deeper facies on platform slopes. The stratigraphic sections were situated in paleogeographically distinct locations, including the eastern margin of the Tethys Sea (Brennecka et al., 2011b; Lau et al., 2016; Elrick et al., 2017; Zhang et al., 2018a), the western margin of the Tethys Sea (Lau et al., 2016; Zhang et al., 2018c), and marginal and seamount sites in the Panthalassa Ocean (Zhang et al., 2018a and 2018b). Because of efforts in characterizing $\delta^{238}U$ in multiple locations, the Late Permian to Middle Triassic record provides an excellent case study regarding the reliability of the $\delta^{238}U$ record hosted in marine carbonates.

The main conclusion of this compilation is that all sections record a rapid and large decrease in $\delta^{238}U$ at the Permian–Triassic boundary, and that the timing and magnitude of this first-order shift are nearly identical in all records despite significant differences in the depositional and diagenetic histories of these

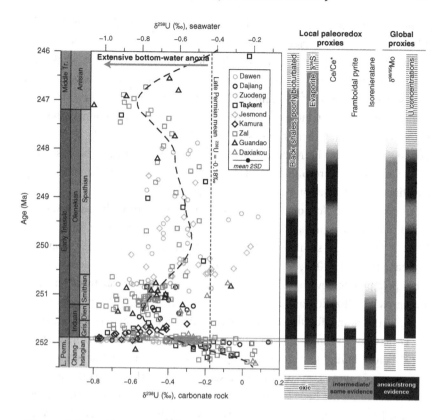

Figure 3 Compilation of δ^{238}U data from the latest Permian to the Middle Triassic. Stratigraphic sections include Dajiang, Dawen, Daxiakou, Guandao, and Zuodeng, located in south China along the eastern margin of the Tethys Sea (circles for shallow and triangles for deeper sites; Brennecka et al., 2011b; Lau et al., 2016; Elrick et al., 2017; Zhang et al., 2018a); Taşkent, Turkey and Zal, Iran from the western margin of the Tethys (squares; Lau et al., 2016; Zhang et al., 2018c); Jesmond, British Columbia, Canada from the eastern margin of the Panthalassa Ocean (Zhang et al., 2018a); and Kamura, Japan, a seamount in Panthalassa (Zhang et al., 2018b; both diamonds). The sections are broadly ordered in the legend from shallowest (platform interior) to deepest (mid-slope) depositional settings. The measured carbonate δ^{238}U values are plotted along the bottom axis while the inferred seawater δ^{238}U, corrected for the ~0.25‰ offset observed in modern carbonate sediments (Chen et al., 2018a; Tissot et al., 2018) is plotted along the upper axis. The dashed line is a LOESS smoothing curve, and the vertical dashed line marks the preextinction mean δ^{238}U of −0.18‰. The horizontal bar marks the extinction interval. The right panel shows inferred interpretations of local and global paleoredox conditions from other proxies, modified from Schaal et al. (2015) and references therein.

samples (Figure 3). The mean $\delta^{238}U$ in all records prior to the extinction is $-0.18‰$ ($n = 67$) and decreases during the main extinction interval to a mean of $-0.46‰$ ($n = 59$) – a statistically significant change (Wilcoxon rank sum test p-value = 9×10^{-11}). The LOESS smoothing curve (black dashed line) shows that carbonate $\delta^{238}U$ remains low through the early Early Triassic, with the overall trend increasing until reaching a local maximum in the late Early Triassic, and then decreasing to $\sim 0.6‰$ at the Early to Middle Triassic transition. The Middle Triassic $\delta^{238}U$ (not shown) has been studied at lower temporal resolution and has a mean $\delta^{238}U$ that is statistically indistinct from the latest Permian (Lau et al., 2016).

As discussed in Section 3.1, investigations of modern carbonate sediment $\delta^{238}U$ indicate that primary carbonate precipitates (corals, algae, ooids, etc.) have $\delta^{238}U$ values similar to that of U in modern open ocean seawater (Romaniello et al., 2013; Chen et al., 2018b), but that sediments derived from these precipitates develop a positive offset of 0.24 to 0.27‰ from seawater during recrystallization and early diagenesis (Chen et al., 2018a; Tissot et al., 2018). Accounting for this offset suggests that the latest Permian seawater $\delta^{238}U$ may have been comparable to or higher than the modern (Elrick et al., 2017) before falling to a mean value of $-0.7‰$ within the extinction horizon (upper axis, Figure 3). A $\sim 0.3‰$ shift in seawater $\delta^{238}U$ (based on differences in the mean values) has been explored using isotopic mass balance models. These are based on tracking changes in both the oceanic U reservoir (N_U) and seawater $\delta^{238}U$, where the output fluxes (J) are commonly simplified to an anoxic sediment sink and a generalized "other" sink with associated isotopic fractionations ($\Delta = \delta^{238}U_{sink} - \delta^{238}U_{sw}$):

$$\frac{dN_U}{dt} = J_{in} - J_{out} = J_{riv} - J_{anoxic} - J_{other} \tag{4.1}$$

$$\frac{d\delta^{238}U}{dt} \times N_U = J_{in} \times \left(\delta^{238}U_{in} - \delta^{238}U_{sw}\right) - J_{out} \times \left(\Delta_{out}\right)$$

$$= J_{riv} \times \left(\delta^{238}U_{riv} - \delta^{238}U_{sw}\right) - J_{anox} \times \left(\Delta_{anox}\right) - J_{other} \times \left(\Delta_{other}\right) \tag{4.2}$$

Model results indicate that variable global redox conditions are the only feasible geologic process that can explain a shift of this magnitude (e.g., Lau et al., 2016), with an estimate of 20 to 60% of the seafloor becoming anoxic immediately after the extinction (Zhang et al., 2018c). Because the Early–Middle Triassic interval has been less intensively sampled, there is less evidence for a negative shift. If supported by additional data, it would suggest a second globally extensive anoxic event that has not yet been captured in other paleoredox proxies (Figure 3; Zhang et al., 2018c).

Another feature of the Permian–Triassic $\delta^{238}U$ compilation is the variability in $\delta^{238}U$ between stratigraphic sections and through time. Some of the stratigraphic trends, particularly in late Early Triassic time, have been interpreted as transient shifts in seawater $\delta^{238}U$ (Zhang et al., 2018a; 2018c), but these patterns are less clear when plotted along the same time axis. However, accurate correlations require well-constrained age models based on conodont biostratigraphy and carbon-isotope chemostratigraphy, which likely introduces error from local variations in $\delta^{13}C$ and in the first and last appearances of rare conodonts. Relatively lower sampling resolutions may also play a role in producing patterns that are less well defined.

Given that modern carbonate sediments also have heterogeneous $\delta^{238}U$ values (Figure 1), some of the variability in the Permian–Triassic record may be explained by heterogeneity produced during early diagenesis. In the modern Bahamas, early and syndepositional diagenesis of U in carbonate sediments produces a spread of $\delta^{238}U$ values, which form a normal distribution with a mean offset of ~0.27 ± 0.17‰ (1 S.D.) between individual drill core samples (Chen et al., 2018a). In deeper Bahamian settings, mixing between calcite (from planktic calcifiers) and aragonite phases has also been suggested to result in variable $\delta^{238}U$ (Tissot et al., 2018), although this cannot explain the Permian–Triassic $\delta^{238}U$ data because there is no systematic difference with water depth and low-magnesium calcite planktic biomineralizers had not yet evolved. Another possible factor is changing seawater U speciation, which may have resulted in a fractionation of +0.18‰ during incorporation of U(VI) into primary $CaCO_3$ precipitates (Chen et al., 2017). Environmental conditions (such as acidification or precipitation of very early marine cements) during Permian to Middle Triassic time may have resulted in geographically variable U speciation as a result of changing seawater carbonate chemistry, possibly contributing to the observed heterogeneity.

In sum, the carbonate $\delta^{238}U$ record spanning the end-Permian extinction and its recovery period is interpreted to track a large expansion of ocean anoxia and can complement other local and globally averaged proxies (Figure 3). Despite the different diagenetic histories of each section, the magnitude and timing of the shift are similar, highlighting the potential of carbonates to track seawater $\delta^{238}U$. This case study also highlights the variability in $\delta^{238}U$ of carbonate rocks: the ~0.3‰ shift captured at the Permian–Triassic boundary may represent the largest shift in the Phanerozoic, and yet the scatter of $\delta^{238}U$ values within a short time interval can be greater than the mean value of the $\delta^{238}U$ shift itself. Given this heterogeneity, multiple records analyzed at high resolution are critical for verifying patterns in seawater $\delta^{238}U$.

4.3 Comparing $\delta^{238}U$ in Carbonate Rocks and Organic-Rich Shales

Because $\delta^{238}U$ in carbonate rocks may track seawater, and $\delta^{238}U$ in organic-rich shales should capture the fractionated U(IV) sink, comparing contemporaneous records of these archives may help to constrain the fractionation between two of the largest output fluxes. Of the published records, the only intervals with contemporaneous carbonate and shale $\delta^{238}U$ data are the modern and Cretaceous OAE2 (Figure 4). During the OAE2 interval (92.75 to 94.48 Ma), the mean carbonate $\delta^{238}U$ of the Eastbourne chalks is –0.43‰ ($n = 63$) and the mean shale $\delta^{238}U$ of a core drilled on the Demerara Rise is 0.02% ($n = 22$), for a difference of 0.45‰ (Montoya-Pino et al., 2010; Clarkson et al., 2018). Although the records overlap in time, a direct comparison is complicated because the records were sampled at different resolutions and are challenging to correlate precisely. In contrast, modern siliciclastic marine sediments deposited in reducing or high-productivity settings have a mean $\delta^{238}U$ of –0.10‰ ($n = 149$; Weyer et al., 2008; Montoya-Pino et al., 2010; Andersen et al., 2014; Holmden et al., 2015; Noordmann et al., 2015; Andersen et al., 2016; Hinojosa et al., 2016; Rolison et al., 2017). This value (which could be further refined in the future by flux-weighting data from each locality) is nearly identical to carbonate sediments deposited in the last 1 Ma that have a mean $\delta^{238}U$ of –0.11‰ following early diagenesis ($n = 92$; Romaniello et al., 2013; Chen et al., 2018a; Tissot et al., 2018).

Modern upwelling zones and partially restricted basins may not be perfect analogues for the Cretaceous OAE2 organic-rich shales, which were deposited on a continental margin exposed to the open ocean (Montoya-Pino et al., 2010). Nevertheless, the difference between the two datasets implies possible differences between the modern and OAE2 U cycles, and more generally for U cycling in ancient oceans. To illustrate, the largest U isotopic fractionation between seawater and sediments in the present is recorded in modern euxinic basins such as the Black Sea. However, water circulation in these settings is typically restricted from the open ocean, and isotopic fractionation during U(VI) reduction results in Rayleigh distillation of the water column isotopic composition (e.g., Rolison et al., 2017). As a result, the $\delta^{238}U$ of dissolved U(VI) in the water column of these basins is up to –0.2‰ lower than that of open ocean seawater. Consequently, the 0.6‰ offset between bottom waters and sediments is present, but the $\delta^{238}U$ of Black Sea sediments is only +0.4‰ heavier than modern open

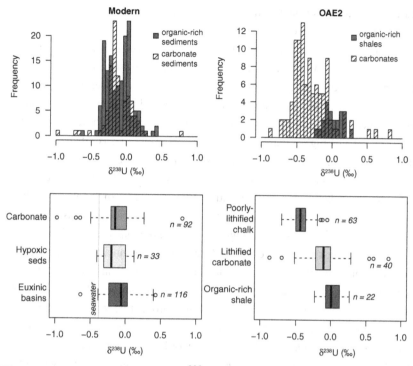

Figure 4 Comparison of carbonate $\delta^{238}U$ data and organic-rich sediments $\delta^{238}U$ data from modern settings (left) and from the Cretaceous OAE2 (right). Histograms of the $\delta^{238}U$ data are shown in the upper panels and box-and-whisker plots are shown in the bottom panels. For the box-and-whisker plots, modern organic-rich sediments are further categorized into euxinic basins and hypoxic sediments, and carbonate data from OAE2 are further delineated into lithified carbonates (South Ferriby and Raia del Pedale sections) and poorly lithified pelagic chalks (Eastbourne section). "Modern" bulk carbonate data are from samples that are <1 Ma in age (Romaniello et al., 2013; Chen et al., 2018a; and Tissot et al., 2018). Modern reducing sediment data are from Weyer et al. (2008), Montoya-Pino et al. (2010), Andersen et al. (2013), Holmden et al. (2015), Noordmann et al. (2015), Andersen et al. (2016), Hinojosa et al. (2016), and Rolison et al. (2017). Carbonate data for OAE2 are from Clarkson et al. (2018), and shale data are from Montoya-Pino et al. (2010).

ocean seawater. The modern continental configuration favors the development of such restricted euxinic basins, but this effect may have been less prevalent in the past. Productive regions in the present oceans (i.e., reducing/hypoxic sediments) do not reach permanently anoxic conditions in the water column, resulting in a smaller positive offset in $\delta^{238}U$. If the modern ocean is not a representative analogue for the past, then constraints from the modern budget will introduce uncertainty in reconstructions of past redox conditions.

In contrast to today, Cretaceous OAE2 is a time defined by widespread anoxia (see references in Montoya-Pino et al., 2010 and Clarkson et al., 2018). The Cenomanian–Turonian Eastbourne chalk is comprised of well-preserved foraminifers and nanofossil grains and was deposited in deeper shelf pelagic environments. In comparison, the $\delta^{238}U$ values of better-lithified carbonates are 0.2 to 0.3‰ higher than at Eastbourne, an effect that is attributed to the diagenetic offset observed in modern carbonate sediments (Clarkson et al., 2018). If the Eastbourne samples preserve seawater $\delta^{238}U$, this would imply a fractionation between seawater and anoxic sediments of 0.45‰, which is comparable to that observed in modern euxinic and restricted basins where the bottom-water $\delta^{238}U$ is known (Figure 1). The potential preservation of this offset may suggest that marine anoxia during the Cenomanian–Turonian transition was not dominated by the restricted basins that are found today. Although speculative, the widespread anoxia of Cretaceous epicontinental seas characterized by numerous OAEs may have been highly productive regions that reached permanently anoxic conditions – comparable to modern oxygen minimum zones, but with greater intensity. In other words, conditions that characterize the Demerara Rise, a nonsilled deep continental margin basin that was influenced by an intense oxygen minimum zone (Montoya-Pino et al., 2010), are not present in today's oceans.

5 Outlook

In this review, we provide an overview of key takeaways from the development and application of the $\delta^{238}U$ proxy and suggest potential future research directions. First, $\delta^{238}U$ has been used to reconstruct redox conditions in multiple sedimentary archives. Agreement with other proxies (for example, during the Great Oxidation Event at 2.4 Ga, the Permian–Triassic event, and Cretaceous OAE2) supports interpretations of $\delta^{238}U$ data as a paleoredox indicator. Second, U isotopes are emerging as a useful proxy in the carbonate rock record, in comparison to the majority of local and global paleoredox indicators that are commonly measured in organic-rich shales. This is an advantage because carbonates have been nearly continuously deposited throughout Earth history and are often found in multiple oceanic basins and spanning multiple facies. Third, due to its redox sensitivity and potential to capture seafloor anoxia, $\delta^{238}U$ can provide unique constraints on oceanographic conditions that are distinct from other proxies. Fourth, $\delta^{238}U$ data from the Archean and Proterozoic may fingerprint the rise of oxidative U mobilization, and potentially changes in atmospheric pO_2, whereas it can provide constraints on marine redox conditions after the atmosphere became oxygenated.

The $\delta^{238}U$ proxy continues to undergo active development. Improving our quantitative estimates of past anoxic conditions is dependent on constraining the isotopic compositions of the sources and sinks of U. As discussed above, ongoing research into understanding the processes and mechanisms leading to fractionation during the diagenesis of carbonate rocks is crucial when using this archive to reconstruct seawater $\delta^{238}U$. Data from modern settings indicate that although the mean offset between carbonates and seawater may be relatively constant, individual sediment samples fall within a wide range of variation. Refining our understanding of this offset and its applicability to past oceans with different carbonate factories and seawater chemistry will require further study. Future investigations could include targeted field studies of diagenesis in modern and ancient environments, as well as experimental and numerical approaches.

Similarly, understanding the fractionation of $^{238}U/^{235}U$ in organic-rich shales, and how this may have varied depending on local depositional conditions, is crucial for characterizing the most powerful lever on seawater $\delta^{238}U$. The variability in modern settings suggests that a wide range of factors can control the fractionation, including detrital composition, productivity, basin restriction, and possibly changes in microbial U reduction. Specifically, potential differences in isotopic fractionation between settings dominated by sulfate reduction (euxinic conditions) versus dissimilatory iron reduction (ferruginous conditions) remain unknown. As discussed in Section 4.3, U isotope systematics of ancient oceans may have been unique. Future research investigating the variability in modern and ancient settings, in addition to diagenetic modeling, may help to improve our calibration of the modern system, which can be further extrapolated to deeper time.

In addition to this summary, we emphasize that studies of intervals where no large redox fluctuations are expected will add new insights into typical variability of carbonate and shale $\delta^{238}U$. Future work should aim to include multiple stratigraphic sections, where possible, to evaluate spatial heterogeneity.

6 Key Papers

Papers That Established the Redox-Sensitive Nature of the $\delta^{238}U$ Proxy

Stirling C. H., Andersen M. B., Potter E.-K. and Halliday A. N. (2007) Low-temperature isotopic fractionation of uranium. *Earth and Planetary Science Letters* **264**, 208–225.

Weyer S., Anbar A. D., Gerdes A., Gordon G. W., Algeo T. J. and Boyle E. A. (2008) Natural fractionation of $^{238}U/^{235}U$. *Geochimica et Cosmochimica Acta* **72**, 345–359.

Brennecka G. A., Wasylenki L. E., Bargar J. R., Weyer S. and Anbar A. D. (2011a) Uranium isotope fractionation during adsorption to Mn-oxyhydroxides. *Environmental Science & Technology* **45**, 1370–1375.

Calibrations of Isotopic Fractionation in Reducing Settings

This paper set up the mechanism by which effective fractionation is subject to local depositional conditions.

Andersen M. B., Romaniello S., Vance D., Little S. H., Herdman R. and Lyons T. W. (2014) A modern framework for the interpretation of ^{238}U/^{235}U in studies of ancient ocean redox. *Earth and Planetary Science Letters* **400**, 184–194.

Calibrations of Isotopic Fractionation and Variability in Carbonate Sediments

Chen X., Romaniello S. J., Herrmann A. D., Hardisty D., Gill B. C. and Anbar A. D. (2018a) Diagenetic effects on uranium isotope fractionation in carbonate sediments from the Bahamas. *Geochimica et Cosmochimica Acta* **237**, 294–311.

Tissot F. L. H., Chen C., Go B. M., Naziemiec M., Healy G., Bekker A., Swart P. K. and Dauphas N. (2018) Controls of eustasy and diagenesis on the ^{238}U/^{235}U of carbonates and evolution of the seawater ^{234}U/^{238}U during the last 1.4 Myr. *Geochimica et Cosmochimica Acta* **242**, 233–265.

Summaries of the U Isotope Budget

Andersen M. B., Vance D., Morford J. L., Bura-Nakić E., Breitenbach S. F. M. and Och L. (2016) Closing in on the marine ^{238}U/^{235}U budget. *Chemical Geology* **420**, 11–22.

Tissot F. L. H. and Dauphas N. (2015) Uranium isotopic compositions of the crust and ocean: age corrections, U budget and global extent of modern anoxia. *Geochimica et Cosmochimica Acta* **167**, 113–143.

Uranium Isotopes as an Indicator of Atmospheric Oxygenation (via Uraninite Oxidation) in Organic-Rich Shale Archives

Kendall B., Brennecka G. A., Weyer S. and Anbar A. D. (2013) Uranium isotope fractionation suggests oxidative uranium mobilization at 2.50 Ga. *Chemical Geology* **362**, 105–114.

Uranium Isotopes in Carbonates

Lau et al. (2016) first showed that δ^{238}U reproduced across multiple stratigraphic sections. Both Clarkson et al. (2018) and Zhang et al. (2018c) compared δ^{238}U trends with other environmental proxies. Clarkson et al. (2018) also demonstrated reproducibility across a depth gradient.

Lau K. V., Maher K., Altiner D., Kelley B. M., Kump L. R., Lehrmann D. J., Silva-Tamayo J. C., Weaver K. L., Yu M. and Payne J. L. (2016) Marine anoxia and delayed Earth system recovery after the end-Permian extinction. *Proceedings of the National Academy of Sciences* **113**, 2360–2365.

Clarkson M. O., Stirling C. H., Jenkyns H. C., Dickson A. J., Porcelli D., Moy C. M., Pogge von Strandmann P. A. E., Cooke I. R. and Lenton T. M. (2018) Uranium isotope evidence for two episodes of deoxygenation during Oceanic Anoxic Event 2. *Proceedings of the National Academy of Sciences* **115**, 2918–2923.

Zhang F., Romaniello S. J., Algeo T. J., Lau K. V., Clapham M. E., Richoz S., Herrmann A. D., Smith H., Horacek M. and Anbar A. D. (2018c) Multiple episodes of extensive marine anoxia linked to global warming and continental weathering following the latest Permian mass extinction. *Science Advances* **4**, e1602921.

References

Abe M., Suzuki T., Fujii Y., Hada M. and Hirao K. (2008) An ab initio molecular orbital study of the nuclear volume effects in uranium isotope fractionations. *The Journal of Chemical Physics* **129**, 164309.

Andersen M. B., Elliott T., Freymuth H., Sims K. W. W., Niu Y. and Kelley K. A. (2015) The terrestrial uranium isotope cycle. *Nature* **517**, 356–359.

Andersen M. B., Stirling C. H. and Weyer S. (2017) Uranium isotope fractionation. *Reviews in Mineralogy and Geochemistry* **82**, 799–850.

Anderson R. F., Fleisher M. Q. and LeHuray A. P. (1989) Concentration, oxidation state, and particulate flux of uranium in the Black Sea. *Geochimica et Cosmochimica Acta* **53**, 2215–2224.

Asael D., Tissot F. L. H., Reinhard C. T., Rouxel O., Dauphas N., Lyons T. W., Ponzevera E., Liorzou C. and Chéron S. (2013) Coupled molybdenum, iron and uranium stable isotopes as oceanic paleoredox proxies during the Paleoproterozoic Shunga Event. *Chemical Geology* **362**, 193–210.

Azmy K., Kendall B., Brand U., Stouge S. and Gordon G. W. (2015) Redox conditions across the Cambrian–Ordovician boundary: elemental and isotopic signatures retained in the GSSP carbonates. *Palaeogeography, Palaeoclimatology, Palaeoecology* **440**, 440–454.

Barnes C. E. and Cochran J. K. (1990) Uranium removal in oceanic sediments and the oceanic U balance. *Earth and Planetary Science Letters* **97**, 94–101.

Bartlett R., Elrick M., Wheeley J. R., Polyak V., Desrochers A. and Asmerom Y. (2018) Abrupt global-ocean anoxia during the Late Ordovician–early Silurian detected using uranium isotopes of marine carbonates. *Proceedings of the National Academy of Sciences* **115**, 5896–5901.

Bigeleisen J. (1996) Nuclear size and shape effects in chemical reactions. Isotope chemistry of the heavy elements. *Journal of the American Chemical Society* **118**, 3676–3680.

Brennecka G. A., Herrmann A. D., Algeo T. J. and Anbar A. D. (2011b) Rapid expansion of oceanic anoxia immediately before the end-Permian mass extinction. *Proceedings of the National Academy of Sciences* **108**, 17631–17634.

Brown S. T., Basu A., Ding X., Christensen J. N. and DePaolo D. J. (2018) Uranium isotope fractionation by abiotic reductive precipitation. *Proceedings of the National Academy of Sciences* **115**(35), 8688–8693.

Bura-Nakić E., Andersen M. B., Archer C., de Souza G. F., Marguš M. and Vance D. (2018) Coupled Mo-U abundances and isotopes in a small marine euxinic basin: constraints on processes in euxinic basins. *Geochimica et Cosmochimica Acta* **222**, 212–229.

Chen X., Romaniello S. J. and Anbar A. D. (2017) Uranium isotope fractionation induced by aqueous speciation: implications for U isotopes in marine CaCO₃ as a paleoredox proxy. *Geochimica et Cosmochimica Acta* **215**, 162–172.

Chen X., Romaniello S. J., Herrmann A. D., Samankassou E. and Anbar A. D. (2018b) Biological effects on uranium isotope fractionation ($^{238}U/^{235}U$) in primary biogenic carbonates. *Geochimica et Cosmochimica Acta* **240**, 1–10.

Clark S. K. and Johnson T. M. (2008) Effective isotopic fractionation factors for solute removal by reactive sediments: a laboratory microcosm and slurry study. *Environmental Science & Technology* **42**, 7850–7855.

Condon D. J., McLean N., Noble S. R. and Bowring S. A. (2010) Isotopic composition ($^{238}U/^{235}U$) of some commonly used uranium reference materials. *Geochimica et Cosmochimica Acta* **74**, 7127–7143.

Dahl T. W., Boyle R. A., Canfield D. E., Connelly J. N., Gill B. C., Lenton T. M. and Bizzarro M. (2014) Uranium isotopes distinguish two geochemically distinct stages during the later Cambrian SPICE event. *Earth and Planetary Science Letters* **401**, 313–326.

Dahl T. W., Connelly J. N., Kouchinsky A., Gill B. C., Månsson S. F. and Bizzarro M. (2017) Reorganisation of Earth's biogeochemical cycles briefly oxygenated the oceans 520 Myr ago. *Geochemical Perspectives Letters* **3**, 210–220.

Dunk R. M., Mills R. A. and Jenkins W. J. (2002) A reevaluation of the oceanic uranium budget for the Holocene. *Chemical Geology* **190**, 45–67.

Elrick M., Polyak V., Algeo T. J., Romaniello S., Asmerom Y., Herrmann A. D., Anbar A. D., Zhao L. and Chen Z.-Q. (2017) Global-ocean redox variation during the middle-late Permian through Early Triassic based on uranium isotope and Th/U trends of marine carbonates. *Geology* **45**, 163–166.

Fujii Y., Nomura M., Okamoto M., Onitsuka H., Kawakami F. and Takeda K. (1989) An anomalous isotope effect of ^{235}U in U(IV)-U(VI) chemical exchange. *Zeitschrift für Naturforschung A* **44**.

Goto, K. T., Anbar A. D., Gordon G. W., Romaniello S. J., Shimoda G., Takaya Y., Tokumaru A., Nozaki T., Suzuki K., Machida S., Hanyu T. and Usui A. (2014) Uranium isotope systematics of ferromanganese crusts in the Pacific Ocean: Implications for the marine $^{238}U/^{235}U$ isotope system. *Geochimica et Cosmochimica Acta* **146**, 43–58.

Henderson G. M. and Anderson R. F. (2003) The U-series toolbox for paleoceanography. *Reviews in Mineralogy and Geochemistry* **52**, 493–531.

Hinojosa J. L., Stirling C. H., Reid M. R., Moy C. M. and Wilson G. S. (2016) Trace metal cycling and $^{238}U/^{235}U$ in New Zealand's fjords: implications for reconstructing global paleoredox conditions in organic-rich sediments. *Geochimica et Cosmochimica Acta* **179**, 89–109.

Holmden C., Amini M. and Francois R. (2015) Uranium isotope fractionation in Saanich Inlet: a modern analog study of a paleoredox tracer. *Geochimica et Cosmochimica Acta* **153**, 202–215.

Hood A. v. S., Planavsky N. J., Wallace M. W. and Wang X. (2018) The effects of diagenesis on geochemical paleoredox proxies in sedimentary carbonates. *Geochimica et Cosmochimica Acta* **232**, 265–287.

Hood A. v. S., Planavsky N. J., Wallace M. W., Wang X., Bellefroid E. J., Gueguen B. and Cole D. B. (2016) Integrated geochemical-petrographic insights from component-selective $\delta^{238}U$ of Cryogenian marine carbonates. *Geology* **44**, 935–938.

Jost A. B., Bachan A., van de Schootbrugge B., Lau K. V., Weaver K. L., Maher K. and Payne J. L. (2017) Uranium isotope evidence for an expansion of marine anoxia during the end-Triassic extinction. *Geochemistry, Geophysics, Geosystems* **18**, 3093–3108.

Kendall B., Komiya T., Lyons T. W., Bates S. M., Gordon G. W., Romaniello S. J., Jiang G., Creaser R. A., Xiao S., McFadden K., Sawaki Y., Tahata M., Shu D., Han J., Li Y., Chu X. and Anbar A. D. (2015) Uranium and molybdenum isotope evidence for an episode of widespread ocean oxygenation during the late Ediacaran Period. *Geochimica et Cosmochimica Acta* **156**, 173–193.

Ku T.-L., Knauss K. G. and Mathieu G. G. (1977) Uranium in open ocean: concentration and isotopic composition. *Deep Sea Research* **24**, 1005–1017.

Lau K. V., Macdonald F. A., Maher K. and Payne J. L. (2017) Uranium isotope evidence for temporary ocean oxygenation in the aftermath of the Sturtian Snowball Earth. *Earth and Planetary Science Letters* **458**, 282–292.

Lu X., Kendall B., Stein H. J., Li C., Hannah J. L., Gordon G. W. and Ebbestad J. O. R. (2017) Marine redox conditions during deposition of Late Ordovician and Early Silurian organic-rich mudrocks in the Siljan ring district, central Sweden. *Chemical Geology* **457**, 75–94.

McManus J., Berelson W. M., Klinkhammer G. P., Hammond D. E. and Holm C. (2005) Authigenic uranium: relationship to oxygen penetration depth and organic carbon rain. *Geochimica et Cosmochimica Acta* **69**, 95–108.

Meece D. E. and Benninger L. K. (1993) The coprecipitation of Pu and other radionuclides with $CaCO_3$. *Geochimica et Cosmochimica Acta* **57**, 1447–1458.

Montoya-Pino C., Weyer S., Anbar A. D., Pross J., Oschmann W., van de Schootbrugge B. and Arz H. W. (2010) Global enhancement of ocean anoxia during Oceanic Anoxic Event 2: a quantitative approach using U isotopes. *Geology* **38**, 315–318.

Morford J. L. and Emerson S. (1999) The geochemistry of redox sensitive trace metals in sediments. *Geochimica et Cosmochimica Acta* **63**, 1735–1750.

Morford J. L., Martin W. R. and Carney C. M. (2009) Uranium diagenesis in sediments underlying bottom waters with high oxygen content. *Geochimica et Cosmochimica Acta* **73**, 2920–2937.

Nomura M., Higuchi N. and Fujii Y. (1996) Mass dependence of uranium isotope effects in the U(IV)–U(VI) exchange reaction. *Journal of the American Chemical Society* **118**, 9127–9130.

Noordmann J., Weyer S., Georg R. B., Jöns S. and Sharma M. (2016) $^{238}U/^{235}U$ isotope ratios of crustal material, rivers and products of hydrothermal alteration: new insights on the oceanic U isotope mass balance. *Isotopes in Environmental and Health Studies* **52**, 141–163.

Noordmann J., Weyer S., Montoya-Pino C., Dellwig O., Neubert N., Eckert S., Paetzel M. and Böttcher M. E. (2015) Uranium and molybdenum isotope systematics in modern euxinic basins: case studies from the central Baltic Sea and the Kyllaren fjord (Norway). *Chemical Geology* **396**, 182–195.

Phan T. T., Gardiner J. B., Capo R. C. and Stewart B. W. (2018) Geochemical and multi-isotopic ($^{87}Sr/^{86}Sr$, $^{143}Nd/^{144}Nd$, $^{238}U/^{235}U$) perspectives of sediment sources, depositional conditions, and diagenesis of the Marcellus Shale, Appalachian Basin, USA. *Geochimica et Cosmochimica Acta* **222**, 187–211.

Rolison J. M., Stirling C. H., Middag R. and Rijkenberg M. J. A. (2017) Uranium stable isotope fractionation in the Black Sea: modern calibration of the $^{238}U/^{235}U$ paleo-redox proxy. *Geochimica et Cosmochimica Acta* **203**, 69–88.

Romaniello S. J., Herrmann A. D. and Anbar A. D. (2013) Uranium concentrations and $^{238}U/^{235}U$ isotope ratios in modern carbonates from the Bahamas: assessing a novel paleoredox proxy. *Chemical Geology* **362**, 305–316.

Russell A. D., Emerson S., Nelson B. K., Erez J. and Lea D. W. (1994) Uranium in foraminiferal calcite as a recorder of seawater uranium concentrations. *Geochimica et Cosmochimica Acta* **58**, 671–681.

Russell A. D., Hönisch B., Spero H. J. and Lea D. W. (2004) Effects of seawater carbonate ion concentration and temperature on shell U, Mg, and Sr in cultured planktonic foraminifera. *Geochimica et Cosmochimica Acta* **68**, 4347–4361.

Schaal E. K., Meyer K. M., Lau K. V., Silva-Tamayo J. C. and Payne J. L. (2015) Oceanic anoxia during the Permian-Triassic. In *Volcanism and Global Environmental Change*, edited by Anja Schmidt, Kirsten Fristad, and Linda Elkins-Tanton, 275–290. Cambridge University Press.

Schauble E. A. (2007) Role of nuclear volume in driving equilibrium stable isotope fractionation of mercury, thallium, and other very heavy elements. *Geochimica et Cosmochimica Acta* **71**, 2170–2189.

Song H-y., Song H-j., Algeo T. J., Tong J., Romaniello S. J., Zhu Y., Chu D., Gong Y. and Anbar A. D. (2017) Uranium and carbon isotopes document global-ocean redox-productivity relationships linked to cooling during the Frasnian-Famennian mass extinction. *Geology* 45, 887–890.

Stylo M., Neubert N., Wang Y., Monga N., Romaniello S. J., Weyer S. and Bernier-Latmani R. (2015) Uranium isotopes fingerprint biotic reduction. *Proceedings of the National Academy of Sciences* **112**, 5619–5624.

Wang X., Planavsky N. J., Hofmann A., Saupe E. E., De Corte B. P., Philippot P., LaLonde S. V., Jemison N. E., Zou H., Ossa F. O., Rybacki K., Alfimova N., Larson M. J., Tsikos H., Fralick P. W., Johnson T. M., Knudsen A. C., Reinhard C. T. and Konhauser K. O. (2018) A Mesoarchean shift in uranium isotope systematics. *Geochimica et Cosmochimica Acta* **238**, 438–452.

Wang X., Planavsky N. J., Reinhard C. T., Hein J. R. and Johnson T. M. (2016) A Cenozoic seawater redox record derived from $^{238}U/^{235}U$ in ferromanganese crusts. *American Journal of Science* **316**, 64–83.

Wei G.-Y., Planavsky N. J., Tarhan L. G., Chen X., Wei W., Li D. and Ling H.-F. (2018) Marine redox fluctuation as a potential trigger for the Cambrian explosion. *Geology* **26**, 587–590.

White D. A., Elrick M., Romaniello S. and Zhang F. (2018) Global seawater redox trends during the Late Devonian mass extinction detected using U isotopes of marine limestones. *Earth and Planetary Science Letters* **503**, 68–77.

Yang S., Kendall B., Lu X., Zhang F. and Zheng W. (2017) Uranium isotope compositions of mid-Proterozoic black shales: evidence for an episode of increased ocean oxygenation at 1.36 Ga and evaluation of the effect of post-depositional hydrothermal fluid flow. *Precambrian Research* **298**, 187–201.

Zhang F., Algeo T. J., Cui Y., Shen J., Song H., Sano H., Rowe H. D. and Anbar A. D. (2018a) Global-ocean redox variations across the Smithian-Spathian boundary linked to concurrent climatic and biotic changes. *Earth-Science Reviews*.

Zhang F., Algeo T. J., Romaniello S. J., Cui Y., Zhao L., Chen Z.-Q. and Anbar A. D. (2018b) Congruent Permian-Triassic $\delta^{238}U$ records at Panthalassic and Tethyan sites: confirmation of global-oceanic anoxia and validation of the U-isotope paleoredox proxy. *Geology* **46**, 327–330.

Zhang F., Romaniello S. J., Algeo T. J., Lau K. V., Clapham M. E., Richoz S., Herrmann A. D., Smith H., Horacek M. and Anbar A. D. (2018c) Multiple episodes of extensive marine anoxia linked to global warming and continental weathering following the latest Permian mass extinction. *Science Advances* **4**, e1602921.

Zhang F., Xiao S., Kendall B., Romaniello S. J., Cui H., Meyer M., Gilleaudeau G. J., Kaufman A. J. and Anbar A. D. (2018d) Extensive marine anoxia during the terminal Ediacaran Period. *Science Advances* **4**, eaan8983.

Acknowledgments

The authors thank M. O. Clarkson for providing raw data. K. V. L. acknowledges support through an Agouron Geobiology Postdoctoral Fellowship, ACS-PRF (Grant 57545-ND2), and the NASA Astrobiology Institute under Cooperative Agreement No. NNA15BB03A issued through the Science Mission Directorate. S. J. R. acknowledges financial support from the US NSF Sedimentary Geology and Paleobiology Program (award EAR-1733598), the NSF Frontiers in Earth System Dynamics program (award EAR-1338810), and the NASA Exobiology Program (no. NNX13AJ71 G and 18-EXO18-0100). F. Z. acknowledges support from a Danish Agency for Science, Technology and Innovation grant (No. DFF 7014–00295).

Geochemical Tracers in Earth System Science

Timothy Lyons
University of California
Timothy Lyons is a Distinguished Professor of Biogeochemistry in the Department of Earth Sciences at the University of California, Riverside. He is an expert in the use of geochemical tracers for applications in astrobiology, geobiology and Earth history. Professor Lyons leads the 'Alternative Earths' team of the NASA Astrobiology Institute and the Alternative Earths Astrobiology Center at UC Riverside.

Alexandra Turchyn
University of Cambridge
Alexandra Turchyn is a University Reader in Biogeochemistry in the Department of Earth Sciences at the University of Cambridge. Her primary research interests are in isotope geochemistry and the application of geochemistry to interrogate modern and past environments.

Chris Reinhard
Georgia Institute of Technology
Chris Reinhard is an Assistant Professor in the Department of Earth and Atmospheric Sciences at the Georgia Institute of Technology. His research focuses on biogeochemistry and paleoclimatology, and he is an Institutional PI on the 'Alternative Earths' team of the NASA Astrobiology Institute.

About the Series
This innovative series provides authoritative, concise overviews of the many novel isotope and elemental systems that can be used as 'proxies' or 'geochemical tracers' to reconstruct past environments over thousands to millions to billions of years – from the evolving chemistry of the atmosphere and oceans to their cause-and-effect relationships with life.
Covering a wide variety of geochemical tracers, the series reviews each method in terms of the geochemical underpinnings, the promises and pitfalls, and the 'state-of-the-art' and future prospects, providing a dynamic reference for graduate students and researchers in geochemistry, astrobiology, paleontology, paleoceanography and paleoclimatology.
The short, timely, broadly accessible papers provide much-needed primers for a wide audience – highlighting the cutting-edge of both new and established proxies as applied to diverse questions about Earth system evolution over wide-ranging time scales.

Cambridge Elements $\equiv$

Geochemical Tracers in Earth System Science

Elements in the Series

The Uranium Isotope Paleoredox Proxy
Kimberly V. Lau et al.

Triple Oxygen Isotopes
Huiming Bao

A full series listing is available at: www.cambridge.org/EESS

Printed in the United States
By Bookmasters